THOMAS EDISON

DISCOVER THE LIFE OF AN INVENTOR

Ann Gaines

Rourke Publishing LLC
Vero Beach, Florida 32964

www.rourkepublishing.com

PHOTO CREDITS:
©Archive Photo, USDI-Edison National Historic Site (Douglas C. Hones/K.Daley)

EDITORIAL SERVICES:
Pamela Schroeder

Library of Congress Cataloging-in-Publication Data

Gaines, Ann.
    Thomas Edison / Ann Gaines.
        p. cm. — (Discover the life of an inventor)
    Includes bibliographical reference and index.
    ISBN 1-58952-122-6
     1. Edison, Thomas A. (Thomas Alva), 1847-1931—Juvenile
    literature. 2. Inventors—United States—Biography—Juvenile
    literature. [1. Edison, Thomas A. (Thomas Alva), 1847-1931. 2.
    Inventors.] I. Title

TK140.E3 G25 2001
621.3'092—dc21
[B]   2001019374

Printed in the USA

TABLE OF CONTENTS

WHO WAS THOMAS ALVA EDISON?

Thomas Alva Edison was the seventh child of Samuel and Nancy Edison. He was born on February 11, 1847, in Milan, Ohio. He attended school for only a few months between the ages of 7 and 12.

At 12 years old, he began selling newspapers on a train. In 1862, he started his own newspaper. Later he worked as a **telegraph** operator. In his spare time he invented better telegraph machines.

A LIFELONG INVENTOR

Before Thomas Alva Edison began his career as an inventor, people who wanted to send a message far away used a telegraph. Edison made telegraphs better. Homes and streets were lit at night by candles and gas flames. He invented electric lights. The **phonograph** and movies were also Edison's inventions. During his life, Edison got 1,093 **patents**.

Edison invented many things in his laboratory.

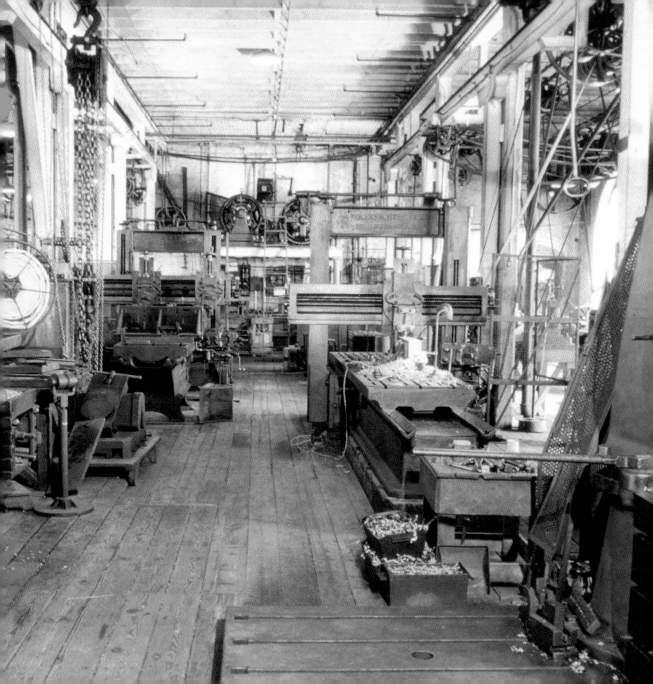

THE PHONOGRAPH

By January 1869 Edison's inventions had made him a lot of money. He supported himself as an inventor. He had made a machine that could send two messages on one wire at the same time.

In 1877, Edison had a new idea. He made a machine called a phonograph. It recorded and let people listen to the human voice. He recorded voices by marking a tinfoil tape.

Edison with his phonograph

Later he ran a needle over the tape. The needle was attached to a speaker. It let people hear what he had recorded.

Edison built the phonograph with John Kreusi's help. It took them 30 hours. The first thing they recorded was the child's nursery rhyme, "Mary Had a Little Lamb." They received a patent for the phonograph in 1877.

This phonograph is a later version of Edison's invention.

THE ELECTRIC LIGHT

On July 29, 1878, Thomas Edison announced he was going to build a safe electric light. His plan was to replace the **gaslights** that were used to light homes.

Francis Upton joined Edison's company in December, 1878. Edison and Upton built an electric light. It was a glass bulb with a **platinum** wire inside. When electric **current** was passed through the wire, the bulb lit. Platinum made the lightbulbs very expensive.

Gaslights and lanterns were used to light homes before Edison's invention.

They tried thousands of different materials to replace the platinum. In October, 1880, they decided **carbon** was best. On December 3, Edison showed the electric lightbulb to other people.

In New York City in January, 1881, the first building was lit by electric lights. Edison's company then opened a power plant in New York City. It supplied electric light for 85 downtown buildings. In 1892, they formed the General Electric Company.

The first lightbulb looked like this.

MOVIES

In 1888, Edison built a movie camera called the **Kinetoscope**. It used a new kind of **Celluloid** film. Edison made these cameras from 1891 to 1896. He sold them mostly to arcades where people came to see films. By 1910, there were more than 20,000 stores that showed films.

The Kinetoscope camera used a new kind of film.

REMEMBERING THOMAS EDISON

Thomas Edison continued to work until the age of 80. Thomas Edison died in West Orange, New Jersey, on October 18, 1931. His inventions are used by all of us almost every day. He made the first talking movie. He invented an electric pen, a copy machine, a microphone, a wireless telephone, and a storage battery. He worked on an electric car, too.

Thomas Edison's home

Today Edison is remembered most for his electric light. His inventions are on display at many museums, including the Smithsonian and the Henry Ford Museum, in Dearborn, Michigan.

Thomas Edison is remembered most for his electric light.

IMPORTANT DATES TO REMEMBER

1847   Born in Milan, Ohio (February 11)

1862   Started his own newspaper

1877   Received patent for phonograph

1878   Edison announced plan for electric light

1881   First building lit by electric light

1886   Married Mina Miller (February 24)

1888   Edison built the first movie camera

1892   General Electric Company formed

1913   Edison made the first talking movie

1931   Died in New Jersey (October 18)

GLOSSARY

carbon (KAR ben) — a very common material in nature that does not cost a lot of money to use

Celluloid (SEL yeh loyd) — motion-picture film made out of a colorless material

current (KUR ent) — flow of electricity through a wire

gaslight (GAS lyt) — lamps that use natural gas

Kinetoscope (ki NET eh skohp) — a machine to make and show movies

patents (PAT nts) — grants made by the government that say only the creator of an invention has the right to make, use, or sell the invention for a period of time

phonograph (FOH neh graf) — a machine that reproduces sound

platinum (PLAT eh nem) — a silver-white material that costs a lot of money

telegraph (TEL eh graf) — a machine that sends messages over electric wires

INDEX

Further Reading

Adler, David A., *A Picture Book of Thomas Alva Edison*. Holiday House, 1996
Linder, Greg. *Thomas Edison: A Photo Illustrated Biography*. Bridgestone, 1999
Wallace, Joseph. *The Light Bulb*. Atheneum, 1999

Websites To Visit

www.si.edu/lemelson/edison/html/thomas_ alva_ edison.html
www.nps.edis/home.html

About The Author

Ann Gaines is the author of many children's nonfiction books. She has also worked as a researcher in the American Civilization Program at the University of Texas.